CW00501261

BRUNEL'S BRIDGES

JOHN CHRISTOPHER

AMBERLEY

Left: W. Heath Robinson's take on the building of the Royal Albert Bridge. Published in the GWR's *Track Topics* in 1935.

About this book
Hopefully this book will encourage you to delve a little deeper into IKB's works, but please note that some sites may be on private property and access might be restricted for reasons of safety and security.

First published 2014

Amberley Publishing
The Hill, Stroud
Gloucestershire, GL5 4EP

www.amberley-books.com

Copyright © John Christopher, 2014

The right of John Christopher
to be identified as the Author of this work
has been asserted in accordance with the
Copyrights, Designs and Patents Act 1988.

ISBN 978 1 4456 3995 6
E-book ISBN 9781 4456 4006 8

All rights reserved. No part of this book may be reprinted or reproduced or utilised in any form or by any electronic, mechanical or other means, now known or hereafter invented, including photocopying and recording, or in any information storage or retrieval system, without the permission in writing from the Publishers.

British Library Cataloguing in Publication Data.
A catalogue record for this book is available from the British Library.

Typeset in 9.5pt on 12pt Celeste.
Typesetting by Amberley Publishing.
Printed in the UK.

Spanning the Years

Isambard Kingdom Brunel and bridges. The connection seems obvious, and none more so than with the Clifton Suspension Bridge. There are several reasons for this. Firstly it was the young engineer's first significant commission and through the Bristol connection it brought about a stream of other commissions, including the Great Western Railway and the first two of his great ships. The second has to be the spectacular location of the Avon Gorge. If the bridge had been built at ground level it would still be a very fine one, but by spanning the gorge it has become a bridge in the sky, a leap of faith that both raises the spirits and inspires.

The bridge at Clifton was the first project into which Brunel poured his heart and soul, and while he might have dismissed some of the plethora of projects that would later come his way, including another suspension bridge – over the Thames in central London no less – it was the Clifton Suspension Bridge that he described as, 'My first child my darling'. But alas, as is so typical of the larger-than-life story of Brunel, it was not to have a happy ending in his lifetime. In 1831 Bristol erupted into riots, partly in opposition to the Reform Bill but also out of resentment of the power held by the city's elite, in particular the Society

Below: Brunel's Royal Albert Bridge over the Tamar, linking Devon and Cornwall. Photographed from the Saltash shore in 2013 during the recent major renovation.

Above: The magnificent Clifton Suspension Bridge over Bristol's Avon Gorge, wearing the new lights that were installed in 2006 to mark the bicentenary of Brunel's birth. See pages 36–37.

of Merchant Venturers. The riots caused a lot of damage to property, but more significantly for Brunel, the ensuing unrest and financial uncertainty caused subscriptions for the bridge to dry up. The foundation stone was laid in August 1836, but by 1843, with the towers standing ready to receive their chains, the money ran out and the work was halted. The chains were sold and the highly conspicuous towers that loomed over the gorge came to be known as 'Brunel's follies'. It was a bitter blow for Brunel who would never see his bridge finished. It was only after his death that through the determination of his colleagues within the Institute of Civil Engineers that the bridge was finished, albeit in a slightly modified form.

The publication of this book celebrates the 150th anniversary of the opening of the Clifton Suspension Bridge on 8 December 1864. It represent a marvellous opportunity to look at Brunel's major bridges as a whole and serves as a reminder of the incredible achievements of Britain's most celebrated engineer.

Brunel's Bridges

Although Brunel is renowned for his work on the Great Western Railway, it is the special 'signature' pieces that really catch our attention. And while the great tunnels, cuttings and embankments are impressive feats they are almost too big or inaccessible to be appreciated. The bridges, however, represent the highlights, those key features where engineer and landscape met head on. Clifton is at the front of the list, of course, but this is also the perfect opportunity to look at his other notable bridges. These are organised by their four main types in terms of construction and materials, beginning with suspension bridges, followed by the brick and masonry bridges, then by iron and concluding with timber.

While the lion's share came about through the need to carry the various railway lines over rivers or roads, not all the bridges were built for the railways. After all, Clifton's is a road bridge, and its close cousin, the Hungerford Bridge, was built for pedestrian use, while the little iron bridge in Bristol's docks – I think of it as Bristol's 'other bridge' – was the direct precursor to the spectacular tubular bridges he built at Chepstow and Saltash. So size in itself isn't a prerequisite and this selection is based on the engineering challenges faced and their importance in the story of Brunel's engineering career.

We are fortunate that so many of Brunel's major bridges have survived, although there have been some casualties along the way, most notably the tubular bridge at Chepstow. But thankfully we still have the Clifton Suspension Bridge, which is in great shape after 150 years of carrying hundreds of thousands of pedestrians and vehicles across the Avon Gorge. And it is here that Brunel's career, and our story, both begin.

Below: Instantly recognisable, the Clifton Suspension Bridge has become the backdrop to thousands of photographs and yet, at the time of Brunel's death, it was little more than a folly, abandoned and unfinished. *(CMcC)*

A Suspended Way

Clifton's suspension bridge represents a leap of faith across the chasm of the Avon Gorge at Bristol. It is a breathtaking monument to the engineering prowess of its creator and, over the 150 years since its completion, the bridge has become the city's most enduring and iconic landmark. The location is everything, as the almost legendary Brunel biographer, L. T. C. Rolt put it:

> Today, in a world satiated with engineering marvels, the grandeur of Clifton Suspension Bridge can still uplift the heart. More surely than any lineaments in stone or pigment, the aspiring sight of its single, splendid span has immortalised the spirit of the man who conceived it ...

The bridge was the first major commission for the young Isambard Kingdom who had newly emerged into the daylight from the shadow of his illustrious father Marc Brunel (later Sir) and, literally, from the darkness of the Thames Tunnel workings in London where he had taken on the role of resident engineer. Commenced in 1825, this was to have been the first major sub-aquatic tunnel and was intended to link Wapping on the north shore of the Thames with Rotherhithe on the south side to provide an additional crossing point for the road traffic pouring to and from the new Surrey Commercial Docks to the south of the river. Its construction had been made possible by Marc's innovative tunnelling shield, which allowed excavation through the London clay and shale to proceed safely by only exposing a small area at a time. That was the theory, but in practice the working conditions were abominable, with the Thames little more than an open sewer, and the tunnel was inundated on two occasions. The first occurred in May 1827, but after Isambard had descended in a diving bell to inspect the damage the hole was eventually plugged and work resumed. But the second flooding, on 12 January 1828, nearly cost Isambard his life. As was his custom he had been working alongside the miners when the torrent of water broke through. While desperately trying to ensure that the others got to safety he was flushed through the tunnel by the torrent of freezing water and was plucked out semi-conscious as it surged up to the lip of the access shaft. Inevitably this brought an end to the project – although only temporarily as it would turn out as the tunnel was finally completed in 1843 – and an end to Isambard's tunnelling. Badly injured in the second flooding he was sent away to recuperate, to Brighton at first, and it would seem that he ended up in Bristol. Although the first written reference to his presence in the city does not appear until an entry in his father's diary in January 1830, it is generally accepted that he had been coming to the city for about two years before that. It may be that he was drawn there by the competition to design a bridge to cross the Avon Gorge, or it may have been down to happenchance. But either way it was the Clifton Bridge that would launch him on his meteoric career as the pre-eminent engineer of his era, and the close connections and contacts he made through the bridge project cemented his associations with the city and directly led to his appointment as the engineer for the Great Western Railway. No wonder that Brunel affectionately refers to the bridge in his journals as, 'My first love'. But as events unravelled, the love affair was to be far from plain sailing.

The first proposals for a crossing over the Avon Gorge had appeared some fifty years or so before Brunel was born. In 1754 a prosperous wine merchant named William Vick got the ball rolling by bequeathing the sum of £1,000, which was to be invested until it grew to £10,000 when it would be used to fund the construction of a stone or masonry bridge. By the start of the nineteenth century Clifton village had become a fashionable place for Bristol's wealthy merchants, and it was the appropriately named William Bridges who published the next proposal for a very grand bridge. This extraordinary structure would have filled the gorge like a stone curtain, its five storeys containing houses, a granary, corn exchange, chapel and a tavern, all of which were to be surmounted by a lighthouse. It was more like a vertical village than a bridge, and inevitably this impossibly expensive scheme was never realised.

By 1829 the Vick legacy had grown to the intended £10,000 mark and

Below: Brunel's No. 3 drawing entered for the first Clifton Suspension Bridge competition in 1829. Note how the cables descend below the deck, a device intended to add stiffness to the deck and borrowed from his father's bridge for the Île de Bourbon.

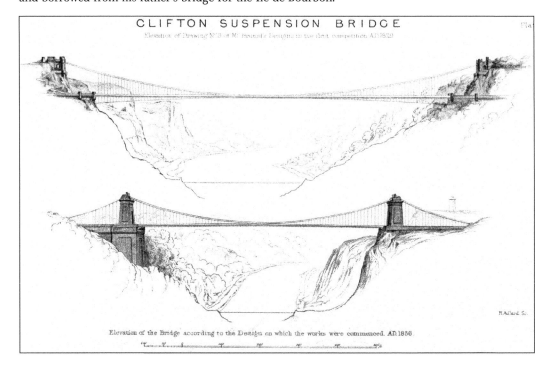

Bristol's merchants decided to push ahead with their plans for a bridge. They quickly discovered, however, that a stone-built one would cost more like £90,000, and taking account of recent advances in bridge construction – in particular the use of wrought iron – they advertised a competition to design a suspension bridge. The young Isambard Kingdom Brunel threw himself into the task and submitted four of the twenty-two entries received by the closing date. His drawings depicted bridges at slightly different sites on the gorge, and featured single spans ranging from 760 feet to 1,180 feet in length. This was far in excess of any suspension bridge then in existence. Telford's Menai Bridge, for example, had a main span of 600 feet while Marc Brunel's suspension bridge on the Île de Bourbon had two spans, each one just under 132 feet.

Of IKB's designs for Clifton, two featured conventional towers perched on the edge of the cliffs, while the others showed shorter towers with the roadway carved through the rocks to incorporate the opening of the existing 'Giants Cave' in the side of the cliff on the Clifton side. With an artistic eye and a sense of theatre Brunel wanted to capitalise on the drama of the location:

> I thought that the effect ... would have formed a work perfectly unique ... the grandeur of which would have been consistent with the situation.

Below: The 'Egyptian thing', the design which was finally accepted by the bridge committee after the second competition. There was some debate as to whether the sphinxes should face outwards to greet the bridge users, or inwards to face each other across the Avon Gorge.

Above: Laying the foundation stone for the Leigh Woods abutment on 27 August 1836. A crowd of 60,000 attended the ceremony, among them Brunel's father, Marc. *Below*: A vision of the future 1830s style. This fanciful illustration shows a Bristol & Exeter Railway train heading through the Ashton Vale on its way to Temple Meads and, in the distance, the Clifton Suspension Bridge, even though this remained unfinished until the 1860s. Brunel was engineer to the B&ER which was built to his broad gauge. Note the carriages carried on the flat-topped wagons – a common practice in the early days of the railways. *(CMcC)*

Correspondence from 1829 has recently come to light which shows how much Isambard had sought his father's advice on the design of the bridge. Marc Brunel was concerned that his son's plans for a single span were too ambitious. He wrote to him saying that it 'needed something in the middle to support' and added, 'you should do it like this,' enclosing a sketch with a towering pagoda splitting the gorge and the span into two. Of course Isambard would have none of it.

The bridge committee invited the seventy-year-old Thomas Telford to judge the entries and he promptly rejected the whole lot. Instead he submitted his own design for a triple-span bridge with two lofty Gothic towers rising to 260 feet from the bottom of the gorge on either bank of the Avon. It was, proclaimed the *Bristol Mercury* in February 1830, a 'singularly beautiful design'. Brunel did not agree and he responded with customary sarcasm:

> As the distance between the rocks was considerably less than what had always been considered as within the limits of which suspension bridges might be carried, the idea of going to the bottom of such a valley for the purpose of raising expense for two intermediate supporters hardly occurred to me.

But he was absolutely right, of course. Telford's towers would have been absurdly costly to build and in the end the bridge committee did what committees do best, they fudged it. While publicly expressing their admiration of Telford's Gothic masterpiece, they privately reopened discussions and in 1830 a second bridge competition was announced. Another ragbag of designs was submitted, including some which entirely ignored the committee's stipulation that it was to be a suspension bridge. William Armstrong, for example, proposed a single-span girder bridge with masonry viaducts to either side, while C. H. Capper put forward a Telford-lookalike with twin towers in a rustic style. Brunel hedged his bets and submitted four designs. The first was a reworking of the Giant's Cave theme. The second and third reduced the span to a more acceptable 720 feet by introducing a masonry abutment protruding from the Leigh Woods side, while the fourth pandered to Telford's prejudices by including a pair of simplified towers rising from the riverbank and dressed in an Egyptian style.

Unbelievably, to Brunel at least, his design came only second in a shortlist of four, and all of the entries were criticised by the judges. Unwilling to accept second place Brunel berated the committee until a modified version of one of his entries was accepted. He referred to it as the 'Egyptian thing' and it featured two towers capped with sphinxes, with the tower on the Leigh Woods side resting on an abutment. In June 1831 a ceremony was held to mark the commencement of construction (and the foundation stone was laid in August 1836). However, later that same year matters took an unexpected turn for the worse when Bristol was rocked by riots. Rioters ransacked the Mansion House in Queen Square and several buildings were razed to the ground. Brunel became personally embroiled in the action when he enrolled as a special

Bristol was a riot in 1831 and Brunel enrolled as a special constable. Prompted by opposition to the Reform Bill and out of resentment of the power held within the city by a small elite, the rioters caused considerable damage in Queen Square. The ensuing unrest and financial uncertainty caused subscriptions for the Clifton Suspension Bridge to dry up.

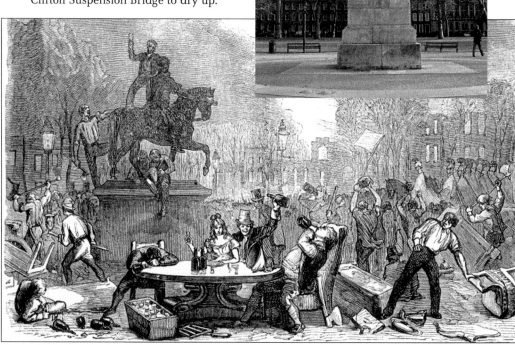

constable. However, the riots had cost the city dearly and the long-term effect was to stem the flow of cash available for speculative projects such as the suspension bridge.

Construction work continued with the existing funds, but by 1843, with the towers standing ready to receive the chains, the money ran out and work was halted. For the next two decades the towers loomed over the Avon Gorge like forlorn gravestones to the unfinished bridge. To Bristolians they were known as 'Brunel's follies'. In 1851 the chains intended for the bridge were bought by the South Devon Railway Company and incorporated within the Royal Albert Bridge at Saltash, which was completed shortly before Brunel died in 1859. At the time of his death it looked as if the bridge would never be completed and there were calls for the towers to be removed. In truth Brunel's reputation was at a low ebb by this time. The Clifton Bridge was uncompleted, his atmospheric railway experiment had been a humiliating failure, the broad gauge railways had been condemned by a Parliamentary Commission in 1846 and two of his great ships, the *Great Western* and the *Great Britain,* had to be sold off after the latter had been driven aground in Dundrum Bay on the Irish coast, also in 1846, thus putting an end to his transatlantic visions for an integrated international transport system. Furthermore the third and final ship, the mammoth *Great Eastern,* had been problematical in its construction and reluctant to launch

Below: The steam packet *Demerara* was caught by the Avon's treacherous tides in 1851. Of particular interest is the unfinished Clifton bridge tower in the background. *(CMcC)*

Left: William Henry Fox Talbot's photograph of the Hungerford Bridge at low tide, looking across the Thames with the south bank and its shot tower in the background.

Above: The Hungerford Bridge opened on 1 May 1845. With a main span of 676 feet it was built as a pedestrian toll bridge linking the south bank of the river with Hungerford Market on the north side. The illustration on the left, *c.* 1850, shows the view of the market as seen from the bridge itself.

– having got stuck on the slipway at Millwall, on the Thames – and its maiden voyage was marred by an explosion that killed two stokers.

There was one other factor. Brunel's other suspension bridge, the lesser-known Hungerford Bridge – a pedestrian crossing over the Thames – was destined to be demolished along with Hungerford Market to make way for a new railway bridge and station at Charing Cross. By chance it was through this unlikely chain of events that Clifton would finally get its bridge.

Hungerford Bridge

By the time he was thirty years old Brunel had amassed an impressive array of engineering commissions which included the design of a footbridge crossing the Thames to connect Lambeth, on the south bank, with Hungerford Market, which was located between The Strand and the northern bank of the river. A market had existed on this site, next to the town house of the Hungerford family since 1682, and was mentioned by the diarist Samuel Pepys. By the early nineteenth century the old market had become dilapidated and it was replaced by a splendid new building designed in the Italiante style by Charles Fowler, who had also been the architect for Covent Garden Market. This opened in 1833 and featured an open court area where fish was sold – with the replacement of the Old London Bridge it became possible for the fishing boats to come right up to the market steps – and fruit and vegetables were on the upper level, with meat sellers on the northern side of a great hall.

It was two years after the new market opened that Brunel was commissioned to construct the suspension bridge to provide a pedestrian crossing. It would appear that this commission had come via his brother-in-law, Sir Benjamin Hawes, a convenient example of nepotism or, as we prefer to call it nowadays, networking. Any other engineer might have seized upon this as an opportunity to create a show-piece of engineering excellence right in the heart of the capital, albeit over a river that would become notorious as the source of the 'Great

Below: Location of the bridge over the Thames. By the time this was published in the 1890s it carried the railway line into Charing Cross station.

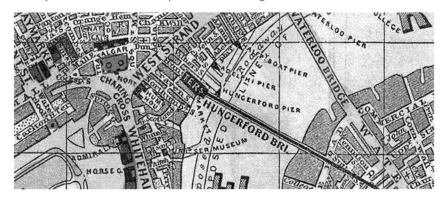

Left: In 1862 Hungerford Market was sold to the South Eastern Railway as a site for Charing Cross station, and work began on demolishing the market buildings.

Above: Work is well in hand with the conversion of Hungerford Bridge to its new use. In this engraving from *The Illustrated London News* the supports are in place and the suspension rods have been detached from the deck.

Left: The artist Claude Monet painted a number of scenes along the Embankment, including this one of the railway bridge in 1903.

Stink' only a few years later. However Brunel's mind was on bigger things by this time and he dismissed the bridge out of hand, as illustrated by his journal entry for 26 December 1835:

> Suspension bridge across the Thames (Hungerford foot-bridge) I have condescended to be engineer of this but I shan't give myself much trouble about it. If it is done it will add to my stock of irons.

In designing this suspension bridge Brunel drew heavily upon his experience gained with the designs for the uncompleted Clifton Bridge. The main difference between the two was that, as the Hungerford Bridge was only intended for pedestrians, its deck would be only 14 feet wide. (Brunel's suggestion that it could be widened to accommodate carriages was not taken up.) Hungerford Bridge had a central span of 676 feet, compared with Clifton's slightly bigger 702 feet, and had two long side spans of 343 feet each. This gave it an overall length of 1,388 feet, which is actually 36 feet more than Clifton's.

The two piers were of hollow construction in red brick dressed with stone. Their extra wide footings were designed to distribute the weight over a wide area, and they served as landing piers for the river steamers with internal staircases leading up to the bridge level. On each pier stood a sturdy tower, in the Italianate campanile style, with four solid pillars of brick to take the weight of the chains. These were supported by saddles which rested on rollers working in oil on a large iron plate. By this means the chains could move horizontally to absorb any deflection caused by uneven loading on the spans, and yet the pressure of the chains on the towers always remained vertical. It also meant that the chains on the land sides could be shorter and at a greater inclination than on the centre span. On either end of the bridge the four chains were anchored within substantial abutments which rested on piles driven at an oblique angle inclined away from the river.

Work on the bridge began in 1841 and in December the following year *The Illustrated London News* reported on progress, stating that the pier on the Surrey side was above the high tide water mark. Under the heading 'Hungerford Suspension Bridge' it provided readers with the following description and outline of the bridge's benefits:

> It is intended for foot passengers only, and promises to be a light, elegant structure, crossing the river with one broad central span from pier to pier. The two piers will be ... rendered light in appearance by architectural adjuncts, but sufficiently substantial to uphold the massive chains, which are estimated to weigh about seven hundred tons ... The general effect of the bridge promises to be good, and likely to improve, very majestically the appearance of Hungerford. Passengers will walk through the centre of the fruit stalls, over the fish market, and in a few minutes find themselves in Pedlar's Acre or – as it has been christened in obedience to the modern fashion for fine

Several views of the bridge and Charing Cross station.

Top left: 1864 engraving from the station looking southwards along the bridge, with the signal box perched on its own gantry above the tracks. Note the shot tower in the background again.

Middle left: The Victoria Embankment, completed in the 1860s, brought the shoreline much closer to the pier on this side of the river. It was built to accommodate the Underground and a new sewage system.

Bottom left: During the Second World War the bridge received a direct hit by a German bomb. Looking towards the station the extended signal box is visible above the tracks.

Opposite: Looking down on the bridge and station. The upper photograph dates from the 1960s, with the stem of the Post Office Tower growing in the distance. The lower one is taken from the London Eye in 2005. Known widely as the Charing Cross Railway Bridge, the Hungerford name came back into use with the addition of the Golden Jubilee Footbridges.

19

THE KEEPER OF THE BRIDGE.

Spirit of Ugliness (to Charing Cross Bridge). "SO LONG AS I'VE ANYTHING TO DO WITH LONDON, YOU SHAN'T GO. YOU'RE MY MASTERPIECE."

Above: During the 1920s there was an unrealised scheme to move Charing Cross station to the south side of the river and rebuild the bridge for road traffic. *Punch* parodied the loss of the rail bridge as a masterpiece of ugliness in this cartoon published in 1923. Every day, thousands of pedestrians use the Golden Jubilee Bridges on either side of the rail bridge.

Above: The wide piers of Brunel's bridge remain, from the deck downwards. They had internal staircases to provide access to river steamers moored there. One of the doorways can be seen on the side. *Below:* The substantial Brunellian abutment on the south side of the river, just along from Festival Hall, with the railway going to Waterloo East station and Tower Bridge beyond.

names – Belvedere-road, Lambeth. Some half mile is expected to be saved, for which the toll is to be a halfpenny.

The bridge opened on 1 May 1845 and despite the stink of the river it proved to be very popular with Londoners, especially following the opening of Waterloo Station on the south side of the river in 1848. It has been suggested that the opening of the bridge might have provided some solace to Brunel to make up for the abandonment of the Clifton bridge, but remaining true to his journal entry he had given the minimum of trouble to Hungerford's design. The result was functional, lacking in any great flair and, inevitably, the setting could never match the drama of the Avon Gorge at Bristol. This may account for the fact that the Hungerford Bridge was so easily forgotten; that and Brunel's lack of interest combined with the extraordinary brevity of its existence.

By the mid-1850s the market was failing financially and matters were made worse in 1854 when the adjoining Hungerford Hall burnt down, causing some damage to the market buildings. It was time for the Hungerford Market Company to sell out and, with a growing demand for the railways to come into central London, Parliament granted permission for the South Eastern Railway (SER) company to build a new station, Charing Cross, on the site. The SER purchased the market in 1862 – three years after IKB's death – and work began on its demolition. On the bridge the chains were removed and the brick towers were dismantled down to the deck level, leaving only the original wide footings. The deck was replaced by a much wider and brutishly functional wrought-iron girder bridge to carry the rails into the station, which opened in 1864. This new bridge was designed by the railway company's engineer, John Hawkshaw (later Sir John), and featured nine lattice girder spans resting on the original piers plus additional pairs of iron columns.

Charing Cross Railway Bridge, as it became known, has attracted a great deal of criticism over the years. It has been described as 'squat, ugly' and 'aesthetically notorious'. Because the railway company was required to maintain a pedestrian crossing, narrow walkways were cantilevered out from either side of the bridge and these continued as paid crossings until the toll was abolished in 1878. Later on, the upstream walkway was incorporated within the railway crossing, leaving only the eastern one for pedestrians and this earned itself a reputation as the haunt of muggers during the twentieth century. Clearly something better was needed, and in the 1990s a competition was held to design new footbridges to run either side of the rail bridge. The winning design, by architect Alex Lifschutz, features slender suspension masts linked by steel rods to the walkways. They were completed in 2002 and although they are officially known as the Golden Jubilee Footbridges, in honour of the Queen's Golden Jubilee, they are more often referred to as the Millennium Bridges. However, I am pleased to say that there is an increasing tendency to use the old Hungerford name both for the footbridges and the railway bridge itself.

Construction of the Clifton Suspension Bridge resumed after Brunel's death. *Left:* The first of the chains is in place. Where they pass over the towers they are not fixed in position, but instead pass over rollers to allow for movement.
Below: The same view, looking towards Leigh Woods, seen today.

It is estimated that seven million people use the twin footbridges to cross over the Thames every year, passing just feet away from what remains of Brunel's forgotten suspension bridge. Beneath the deck the brick and masonry piers are clear to see, although the curved red-brick and stone embellishments on the top of the piers are later additions. Curiously, the northern or Middlesex pier appears to be much closer to the shore than the other one, and this is because the northern riverbank was extended with the construction of the Victoria Embankment in the 1860s to accommodate a new sewage system – to deal with that notorious stink – and tunnels for the Metropolitan District underground railway.

Completing the Clifton Bridge

There is one last chapter to the Hungerford Bridge story, one last link between the cities of London and Bristol. Following Brunel's death in 1859, his fellow engineers at the Institute of Civil Engineers decided to complete the Clifton bridge as a monument to their late friend and colleague. John Hawkshaw – who happened to be the engineer behind the new Charing Cross railway bridge being constructed on the piers of the old Hungerford Bridge – together with W. H. Barlow, wrote a feasibility study showing how the Hungerford chains could be used to complete the Clifton Bridge. The wrought-iron chains which had been originally made for Clifton, but never used there, had already been recycled by Brunel himself for the Royal Albert Bridge at Saltash. So Hawkshaw purchased the second-hand set of Hungerford chains for £5,000 and took them to Bristol where, augmented by an additional set to carry the greater load, they are now draped across the Avon Gorge.

In order to complete the Clifton Suspension Bridge several significant changes were made to Brunel's design. Firstly the sphinxes and the fine Egyptian decoration on the towers were out. The deck of the bridge was widened from 24 feet to 30 feet. For the deck Brunel had specified a timber framework with iron straps, but instead it was constructed with wrought-iron plate girders and lattice cross girder. In addition Barlow added two longitudinal wrought-iron girders to support it and provide the main strength. These are visible as the dividers between the roadway and the footpath. To construct it, 16-foot sections were lowered into place by cranes and the suspension rods were then attached, 162 in all, varying in length from 3 feet at the centre of the bridge to 65 feet at the towers. The completed bridge was load-tested with 500 tons of stone ballast and it was found that the middle section sagged just 7 inches, which was well within acceptable tolerances. Pedestrians will notice that the bridge is constantly moving up and down with the weight of the vehicles. It also moves sideways in strong winds, and in higher temperatures the longer rods can expand by up to 3 inches. The road level of the bridge is slightly higher on the Clifton side, by 3 feet in fact, and Brunel deliberately incorporated this into the design to create the appearance that the bridge is level.

In addition the suspension chains from the Hungerford Bridge were

Above: The opening ceremony on 8 December 1864. *Below:* Detail of the tower on the Clifton side with Egyptian-style capping, a nod to Brunel's 'Egyptian thing'.

Above: The sweeping curve of the suspension chains. Note how the deck curves upwards slightly towards the centre of the bridge. *Below:* Egyptian-styled gatehouse on the Leigh Woods side.

S.S. FAITHFUL.   RIVER AVON.
BRISTOL. 88.

Garratt

28

Spanning the spectacular Avon Gorge, the bridge has become an enduring symbol of the city, and naturally it has appeared on countless postcards over the years. *Opposite top:* This photograph from around 1910 was taken from Rownham Ferry at high tide. *Lower:* The SS *Faithful*, looking towards Bristol from the Leigh Woods bank. *This page, top:* A Bristol Channel paddle steamer photographed at Hotwells. During the war many of these paddlers were requisitioned by the Royal Navy, often for use on mine-clearing duties, hence the uniform battleship grey paint job. *Bottom:* View showing the Rownham, later Clifton Bridge, station. *(CMcC)*

strengthened with the addition of a third chain, and they are made up of sections with alternately ten or twelve links of flat metal placed side by side and bolted together. At either side the chains run over the towers and down to ground level where they are deflected into excavations 60 feet deep into the rock. Originally they were anchored underground by a system of Staffordshire brick infills which spread them out like a wedge. However, since the 1920s they have been anchored by 20 feet of solid concrete

Work on completing the bridge recommenced in 1863 and it was only eighteen months later, on 8 December 1864, that it was officially opened to a cacophony of brass bands, cheering crowds, church bells and the firing of a field gun salute.

Since 1952 the bridge has been looked after by a trust. To cover its upkeep, car drivers pay a modest toll, while pedestrians and cyclists have crossed for free since 1991. In 2006 Bristol became the focus of the Brunel 200 celebrations which included a number of events in the city. The Clifton Suspension Bridge was treated to a new lighting system with 3,000 light-emitting diodes installed to illuminate the chains, plus low-level floodlighting for the towers and abutments. On the night of 9 April 2006, the bicentenary of his birth, the bridge became the perfect launchpad for a spectacular fireworks display that lit up the gorge with chains of light and colour.

A Hidden History

When you visit the bridge you can't fail to notice the prominent signs for the Samaritans organisation, after all that is what they are there for. Sadly the bridge has a darker side and since its opening this high structure has attracted many unfortunate visitors with suicide in mind. Nowadays the side railings are much higher than they used to be and the bridge staff are always on the

Opposite: The classic view of the bridge from the entrance to the Cumberland Basin, and the scene on 19 July 1970 when Brunel's iron ship, the *Great Britain*, passed under the bridge, for the only time, on her return to Bristol. *Below:* A memorial plaque on the Clifton approach.

lookout for any tell-tale signs among the stream of pedestrians. Remarkably, there have also been some extraordinary close escapes over the years. In 1885 a young woman named Sarah Ann Henly jumped from the bridge, but her fall was arrested by her wide skirts, which billowed out like a parachute, carrying her relatively gently down to the water. Then in 1896 two girls, Elsie and Ruby Brown, were hurled from the bridge by their father who was in a state of 'temporary derangement' according to the contemporary reports. Whether it was their young age, or the high tide, or possibly a combination of the two, the girls survived and were rescued by the pilot boat. There is also the curious case of Lawrence Donovan who contrived to fake his fall using a dummy and helpers who obligingly shouted, 'He's jumped,' as the dummy was tipped over the side. Their associates made a big show of pulling the real Donovan out of the water. He later claimed that he had survived the fall because he was protected by metal plates which produced protective electricity.

The portal offered by the bridge and gorge has also exerted an irresisible attraction to a number of aviators, and a deadly one to some. The first to go through was a Frenchman, M. Tetard, in 1911. He survived, but when Flying Officer J. G. Crossley of 501 Squadron tried it in 1957 he was killed when his Vampire jet smashed into the side of the gorge at 450 mph. More recently Bristol has become a centre for hot-air ballooning and the occasional aeronaut has been known to take on the challenge even though it is illegal for any aircraft.

The other aspect of the bridge's hidden history came to light in 2002 when it was discovered by chance that the Leigh Woods abutment is actually hollow. Much was made of this dicovery at the time, but I had always assumed that this was the case as with Brunel's great brick-built viaducts, such as the one at Wharncliffe which is described in the section on brick bridges. To infill such a big structure would have been a massive and entirely unnecessary task. In 2005 I had the opportunity to put this to the Bridgemaster at that time, John Mitchell, who I interviewed in connection with the Brunel 200 anniversary. He was inclined to agree:

I thought it must have been built in the same way as a viaduct. Ever since it was built the face of the abutment has suffered from lime staining, indicating that water was carrying free lime. But we had no drawings. In the 1960s Sir Alfred Pugsley, who was a trustee, carried out several bore tests and found nothing. And when Sir Alfred says the abutment is solid, then it's solid! Incredibly the borings had missed their target, as did more recent probing by ground radar. Small internal chambers had been found on the Clifton side in 1978 and again in 1999. but it wasn't until a 3-foot shaft was uncovered during some paving repairs that the truth beneath the Leigh Woods abutment was finally revealed. We assumed the shaft was to do with drainage and so we erected a tripod above it and one of our abseiling specialists, John Corber, went down to take some measurements. He was being lowered when I heard a sudden cry and a few choice words! I immediately looked into the hole, but he had gone and the rope was slack. Naturally I feared the worst. Then his feet suddenly reappeared and he shouted up that he had just put his head into a chamber 36 feet high!

There turned out to be twelve chambers in all with seven on an upper level and five more below, all linked by a maze of narrow shafts. The biggest is almost cathedral-like in its proportions: 60 feet long and 36 feet high. Each chamber is in complete darkness, and stalactites of limestone stretch down almost to the ground. Barely a finger's width these hollow strands, known as 'candles', are very delicate and therefore public access is not possible.

And finally, if your appetite for curiosities remains unsated, then see if you can explain the differences between the two towers. The Clifton one has a much less pointed upper arch, its corners are square, whereas the other tower has rounded-off corners, and it is the only one to have side slits or apertures. Furthermore, nobody knows why there are several courses of lighter coloured stone on the Leigh Woods tower.

The Clifton Suspension Bridge is littered with inscriptions and plaques and there is one up high on the Leigh Woods tower in Latin: 'Suspensa Vix Via Fit'. It is a pun on the name of William Vick and roughly translates as, 'A suspended way made with difficulty'.

Opposite: The great thing about taking a walk in the sky on the suspension bridge is that it is free to pedestrians. The visitor centre is on the Leigh Woods side.

The Clifton Suspension Bridge has always been illuminated, at first with magnesium flares, although these tended to blow out, and since the 1930s by electric lights. The chain bulbs, shown above, were added to mark the Queen's Coronation in 1953. For the bicentenary of Brunel's birth, in 2006, the lighting was upgraded with 3,000 light-emitting diodes (LEDs) on the chains, plus flourescent lighting for the walkways and additional low-level flood lighting on the towers and abutments.

Brick and Stone

Although Brunel is rightly famed for his spectacular suspension bridges and his innovative use of wrought iron, he was also an exponent of bridge-building with more traditional materials such as brick, stone and, as we shall see later, timber. His use of brick is especially interesting as, in a typically Brunellian manner, he would push this most basic building material right to the limit. Sometimes beyond.

The main application for brick-built bridges was on the railways, in particular the early ones, and the Great Western Railway between London and Bristol is blessed with some superb examples of both wide and sometimes oblique brick structures. All of the examples cited here involve river crossings, which is why the Wharncliffe Viaduct is also included under the definition of a bridge. Wharncliffe is also our natural starting point on the GWR main line going from London and heading westwards towards Bristol.

Warncliffe Viaduct

The valley at Hanwell was the first major physical obstacle to be crossed by the railway departing from London. Located almost immediately to the west of Hanwell station, the valley is wide, deep and bisected on the eastern side by the comparatively narrow River Brent. To cross it Brunel constructed the Wharncliffe Viaduct, which at 896 feet long and 65 feet high was the largest brick-built structure on the GWR. It consists of eight semi-elliptical arches of 700-foot span, mounted on elegantly tapered piers capped with sandstone cornices which evoke the Egyptian styling he had previously proposed for the Clifton Suspension Bridge. Apart from providing decorative detailing, these stone cappings served as the supports for the wooden formers or 'centerings' which held the brick arches during construction until the mortar had fully set. The brick piers themselves are hollow, to make the structure as light as possible, and below ground the foundations descend to the London blue clay, with the base of each pier having a footprint of 252 square feet.

The viaduct is named after Lord Wharncliffe, the chairman of the House of Lords committee who supported the building of the railway and had been instrumental in smoothing its passage through Parliament. His coat of arms can be seen high on the southern side of the brickwork. J. C. Bourne's etching of the viaduct despicts a self-assured structure striding through an otherwise rural landscape. Today the urban sprawl is snapping at the heels of this giant, although the immediate area of the valley has been preserved as a public park and you get an unobstructed view of the viaduct from the A4020 Uxbridge Road. The main change to the viaduct since Brunel's time was the addition of the third piers to accommodate the widening of the track in the 1870s. Only a

The Wharncliffe
Viaduct at Hanwell.
It has eight semi-
elliptical arches
supported on brick
piers. Note the coat
of arms of Lord
Wharncliffe, a
supporter of the GWR
Bill in Parliament,
above the central pier.

Left: J. C. Bourne's
highly romanticised
depiction of the
viaduct which was
published in 1846.
It shows an almost
rural scene with the
Egyptian styling
suggestive of ancient
cities. When the
track was widened
in the 1870s, a third
matching pier was
added on the north
side. In this close-up
of the brick piers,
lower left, you can see
the join to the right of
the central pier.

single extra pier was needed as the widening incorporated two extra lines of the narrower 'standard gauge' whereas the original double spans had been for two lines of Brunel's broad gauge track. This is a common feature on the other bridges we shall encounter on this section of the widened track. The widening of the Wharncliffe Viaduct closely matches the original style and is barely noticeable apart from the join line running its full length on the northern side of the central pier.

The crossing over the Uxbridge Road is just to the west of the viaduct, and with Brunel's use of both stone columns and iron girders in the construction of the original bridge, it is covered in the section on his 'Iron' bridges.

Above: GWR illustration of the Cheltenham Flyer crossing Maidenhead's bridge.

Maidenhead Bridge

In 1838 the first section of the GWR to enter service went only as far as Taplow station on the eastern side of Maidenhead (the newer Maidenhead station being constructed about thirty years later). This situation was to last until July the following year and it was caused by the construction of a bridge to cross the 300-foot-wide Thames on that side of the town. It was here that Brunel built his celebrated Maidenhead Bridge, which features two of the widest and flattest arches ever carried out in brick. With a central pier standing on a shoal mid-river, the two main semi-elliptical arches each have a span of 128 feet with a rise of just 24 feet 6 inches. Four smaller semicircular arches join with the

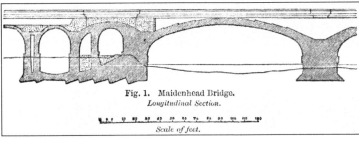

Fig. 1. Maidenhead Bridge.
Longitudinal Section.

Scale of feet.

Left: Sectional side view of the Maidenhead bridge showing the hollow nature of the brickwork structure. Each main span was 128 feet wide.

To cross the 300-foot-wide Thames at Maidenhead, Brunel devised this brick-built bridge consisting of two main spans and a central pier standing on a midstream shoal. With each span 128 feet wide with a rise of just 24 feet 6 inches, the sceptics said it would fall down once the wooden supports, known as centerings, were removed. They were wrong, of course, and the bridge now supports the weight of the much heavier modern trains.

Bourne's view of the bridge is shown at the top of the opposite page.

This page: The view from the eastern bank, with the A4 road bridge visible in the distance. There is a footpath on this side of the river. The almost finished bridge, with centerings still in position, is shown middle right. Plus a commemorative plaque on the eastern side.

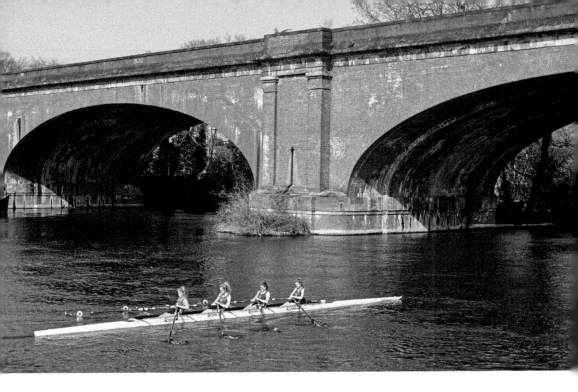

embankment to either side. The result has been described as 'visually the most pleasing and technically the most daring of Brunel's designs in brickwork'. In the biography of his father, Isambard Brunel (junior) wrote:

> The Maidenhead bridge is remarkable not only for the boldness and ingenuity of its design, but also for the gracefulness of its appearance. If Mr Brunel had erected this bridge at a later period, he would probably have employed timber or iron; but it cannot be a matter of regret that this part of the Thames, although subjected to the dreaded invasion of the railway, has been crossed by a structure which enhances the beauty of the scenery.

But at the time of its construction the design of the bridge with its wide flat arches was regarded with suspicion and much derided by some experts who confidently predicted that the bridge arches would not stand at all once the wooden centerings, sometimes known as 'falsework', were removed, let alone bear the weight of the trains. The 1935 GWR publication *Track Topics*, subtitled 'A Book of Railway Engineering for Boys of All Ages', takes up the story:

> There was doubtless joy in the hearts of those experts when, on the centerings of the bridge being removed, the eastern arch disclosed a slight distortion. It was actually a separation of about an inch in the lowest three courses of bricks, due to the centering being eased before the cement had time to properly set. The contractor admitted that he alone was to blame and the fault was duly remedied.

Once this remedial work had been completed, Brunel ordered that the timber centerings should be eased but left in position over the winter. However, when a severe storm in the autumn of 1838 blew them away the bridge was left standing as straight and true as it does to this day. As with the Wharncliffe Viaduct, the bridge at Maidenhead was also widened in the 1870s and it is a tribute to the engineer that this busy stretch of line copes with trains of a weight that would have seemed inconceivable when it was built.

Located to the eastern side of the town, and visible from the A4 road bridge, the Maidenhead Railway Bridge happens to be the most easily accessed of the Thames brick-built bridges for the Brunel-hunters.

Upper Basildon and Moulsford

Continuing westwards and following the Thames Valley, the London–Bristol line crosses the Thames in two places, in quick succession, carried aloft by a pair of fine brickwork bridges. Both of these are on a skew. At Upper Basildon the Thames bends to the west and the bridge crosses into Oxfordshire at an angle of 15 degrees, while 2 miles away the Moulsford Bridge re-crosses the river at a much sharper 45 degrees. This makes the Moulsford Bridge the more interesting of the two and by virtue of a footpath from the A329 Wallingford Road, just beyond the Moulsford Preparatory School, it is the more accessible. Both feature elliptical barrel-arches with face rings, cornices and copings of creamy sandstone. Standing under the Moulsford arches you can appreciate

Below: J. C. Bourne's lithograph of the brick-built bridge over the Thames at Upper Basildon.

The bridge at Moulsford is far more accessible than its counterpart at Upper Basildon. It is shown, opposite, with a First Great Western HST 125.

the complicated compound curve of the brickwork. When it came to the doubling of the line a second parallel bridge was built alongside the original and separated by a narrow gap.

In comparison with the high-profile bridges, such as Wharncliffe and Maidenhead, and given their more rural locations, the Upper Basildon and Moulsford bridges have not attracted the same degree of interest from the historians of the railway. However, J. C. Bourne provides the following description in his volume of lithographic prints, *The History and Description of the Great Western Railway*, published in 1846:

> These bridges are each composed of four elliptical arches of 62 feet span. They are of red brick, in part from the brickfields of Upper Basildon, where are found traces of the plastic clay, and in part from the local loam bed found upon the chalk near Moulsford. The cutwaters, face-rings of the arches, the cornices and copings, are of Bath stone. The spandral walls support longitudinal barrel arches of brick, on which the ballast of the road-way [railway] is laid.

Of the two bridges, J. C. Bourne produced a print of only the Upper Basildon one, shown on page 45. Beyond Moulsford the Thames turns northwards towards Oxford while the railway turns to the west. It is not until it nears Bath that it once again crosses a major river, the Avon, and since it was a long way from the brickfields, Brunel turned to the local stone for his next bridges.

Above: The stone bridge at Bathford, just to the east of Bath itself.

Crossing the River Avon at Bathford ...

The first crossing over the River Avon, as the railway threads its way down the valley towards Bath, is at Bathford. This is probably one of the least known of Brunel's bridges, mainly because of its location but also because it is overshadowed by the presence of Bath almost immediately to the west. As George Bradshaw put it in his celebrated railway guide of 1863:

> The hamlets of Bathford, Bathampton, and Batheaston are now passed in rapid succession, and swerving slightly to the south, the outskirts of the 'Stone-built-city' rise in all their magnificence before us, as if evoked by a magician from the fertile pastures we have so recently quitted.

At least J. C. Bourne thought the stone bridge at Bathford was worthy of attention, and he comments that although 'perfectly plain' it is, nonetheless:

> ... one of the most beautiful structures on the line; it is a bridge of one arch, elliptic, of 54 feet span, and 27 feet rise; the arch is flanked by two projecting piers, and meets the embankment with low long walls; above the whole is a plain entableture and parapet.

48

... and at Bath

The line then enters Bath and passes through Sydney Gardens – the 'Vauxhall of the place' says Bourne – in one of the most elegant railway cuttings to be found anywhere. It then disappears underneath Sydney Road and, passing through two short tunnels under the houses, it then crosses Pulteney Road (the A36) before being carried on the thirty-one arches of the curving Dolemeads viaduct which leads to Bath Spa station nestling within the elbow of the Avon.

Approaching from the eastern side, the masonry bridge over the river is St James's which has a single 88-foot main arch plus two side or 'road' arches. Built of the Bath stone, over the years it is has become peppered with a patchwork of mostly blue-brick repairs. At first sight St James's Bridge appears to be quite plain. However, it has been blessed with some decorative features, albeit very subtle ones. Note in particular the way the cornice of the parapet rests on worked corbels on either side. St James's is a very pleasing structure and the high arch over the river explains why the platform level at Bath is raised up to what appears to be the first floor of the station building. The end of the platform does provide a very good vantage point from which to view the bridge at rail level. The riverside footpath provides good access and passes by the entrance from the river into the Kennet & Avon Canal.

Photographed in 2005, an HST 125 crosses St James's Bridge on the approach into Bath station.

Top left: Sculptural detail on the St James's Bridge, which is located immediately on the east side of Bath station.

Middle left: J. C. Bourne gives us this view of the newly completed bridge as seen from the northern side of the line. The station is off the picture on the right-hand side. It is pretty much impossible to replicate this view nowadays because of the later buildings at this location.

Below: Looking down upon Bath, with the railway line from London coming into the station over the river on the right, and the site of the skew wooden bridge, and the start of the western viaduct, going off to the left. The station is on an awkward site, nestled within the bend of the River Avon.

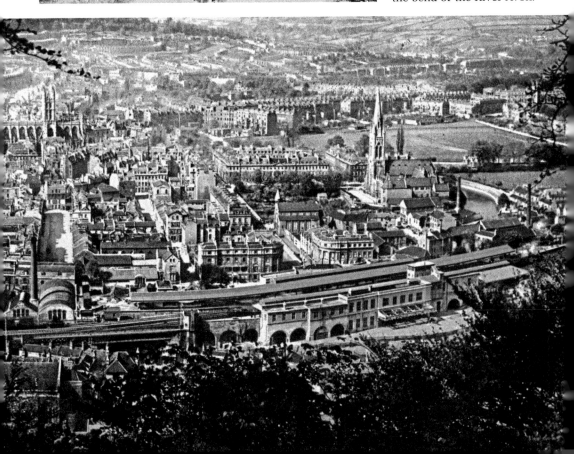

Bath's Viaducts

On the other side of Bath station the railway immediately crosses the river again and Brunel's timber bridge for this site is described in the final section of this book. Heading westwards the line is carried on two long viaducts. The first begins immediately after the river crossing and consists of seventy-three arches; its most notable feature is the castellated section opposite Southgate, in central Bath, which incorporates two road crossings which make up a large oval-shaped roundabout linking the A36 Lower Bristol Road with the A367 Wells Road that heads off to the south. Within this roundabout the central section features two mock towers, in the Gothic style, with the London

Below: The fine crenellated viaduct on the A36 roundabout to the west of Bath station. Note the coats of arms of the GWR and the locomotive wheel. The wide arches originally featured Brunel's much-favoured Elizabethan/Gothic styling.

Typically Brunellian road bridge over the railway line, located at the western end of Oldfield Park station, which opened in 1929. *Below:* The station at Twerton is perched on the side of the line, beside the A36 out of Bath. Opened in 1840 it was closed in 1917 as an economy measure and never reopened. The dwellings created under the Twerton Viaduct are shown opposite.

and Bristol coats of arms in the centre above a locomotive wheel. The road crossings originally had the Brunellian flat-topped Tudor type of arch, although these were replaced with the present steel girders as part of the rebuilding in 1911. The line continues on the viaduct until a low cutting at Oldfield. The Oldfield Park station is post-Brunel and was opened by the GWR in 1929. It does, however, provide an excellent vantage point to get a close-up view of the Brook Road bridge over the line, with another Tudor-style arch, and the similar Bellots Road bridge can be seen in the distance.

After the Oldfield cutting the line takes to another viaduct which more or less follows to the north of the Lower Bristol Road as it leads out of Bath. This is the Twerton Viaduct consisting of twenty-eight arches and a number of road crossings, most notably at Mill Lane, the appropriately named Connection Road, and, opposite MacDonalds, the sharply skewed crossing over the High Street. This is of interest as it is immediately next to the old Twerton station which is accessed from the road level, on the A36 side, and rises up to perch beside the railway line. Twerton station opened in December 1840, and was renamed as Twerton on Avon in 1899 to avoid confusion with Tiverton station. Following closure in 1917, as an economy measure, it never reopened. The building has stood empty for many years although some minor repairs have been undertaken to keep the weather and the squatters out.

To the west of the old station at Twerton you can see the remains of Brunel's attempt to create homes in the arches within the viaduct. This was not an entirely original concept and similar accommodation had been created on other railways including, for example, beneath the elevated London & Blackwall Railway. It is thought that Brunel created these archway dwellings at Twerton to make up for the many cottages that were demolished in the building of the viaduct. However, the two-room accommodation, which only had windows onto the road-side of the viaduct, were not at all popular for obvious reasons.

A Final Avon Crossing at Bristol

This is Brunel's hidden bridge, a Grade I listed masonry structure passing over the River Avon on the approaches to Bristol's Temple Meads station. It has a wide central arch and two smaller flanking arches in Brunel's much-favoured Elizabethan Gothic style.

Unfortunately it is almost impossible to see the bridge nowadays as it is hemmed in on both sides by later steel girder bridges which were added to carry additional lines. It is even difficult to get anywhere near to the bridge as the immediate area has been built on and is occupied by a variety of commercial and light industrial units. The best view is at a distance from the Feeder Road bridge. Thankfully Bourne comes to the rescue and the bridge features in two prints, both as a complete view and also within the wider landscape of Bristol and the surrounding area. The centre arch, he informs us, has a span of 100 feet.

Ironically, the good news is that at least the masonry bridge has survived rather than been completely replaced. One of the additional girder bridges is no longer in use and perhaps one day Brunel's 'hidden' Bristol bridge over the Avon might be revealed once more.

Opposite: Contrasting views of Brunel's masonry bridge over the River Avon on the approach into Temple Meads station. As depicted by Bourne this was a fine bridge, and although it is still there it is now hemmed in on either side by later lattice girder bridges. *Below:* Looking towards Bristol from the eastern side, with the railway line coming in from the right to cross the River Avon via the bridge on its way into Temple Meads. J. C. Bourne, 1846.

Above: The Somerset Bridge over the Parrett was to have been another brick-built structure.

A Bridge Too Wide

All of the bridges covered so far have been on the GWR main line between London and Bristol, but Brunel's network of broad gauge rails extended throughout the South West, through Somerset, Devon and into Cornwall, and also across South Wales and up towards Oxford, Worcester and the West Midlands. The major bridges on these routes are covered in later sections under the relevant construction types – iron and timber – but this round-up of the brick and masonry bridges would not be complete without looking at the brick bridge that failed.

In many cases with the additional railway lines that came Brunel's way following on from the GWR, and to a great extent because they were either extensions of that line or directly connected with it, he acted more in the role of consulting engineer – a term he abhorred – with an extensive team of assistants undertaking the legwork. Those he trusted were given a suprising degree of autonomy and responsibility, with Brunel retaining overall control of the individual projects. One such case was with the Bristol & Exeter Railway, which shared the GWR site at Bristol Temple Meads station, although not directly connected by rail until later, and was built to the master's broad gauge. Brunel appointed William Gravatt, a trusted assistant from the Thames Tunnel days, to survey the route and he was also made resident engineer of the line going from Bristol through Somerset as far as White Ball on the Somerset/Devon border. The major obstacle on the Somerset stretch of line was the River Parrett to the south of Bridgwater. Here Gravatt proposed another wide-spanned brick bridge, and at 100 feet wide and with a rise of just 12 feet this Somerset Bridge was even flatter than the one at Maidenhead. It is not known how much involvement Brunel had in the design, but certainly, at the very least, he must have approved the drawings. Work started on the Somerset Bridge in 1838 and the line opened as far as Taunton on 1 July 1842. As with Maidenhead the centerings of the Somerset Bridge were left in place, but there were problems with the masonry foundations settling into the softer ground. Brunel was forced to remove them in 1843, to provide a passage for vessels on the river, but with the foundations pushing outwards a wooden structure was added within six months. The present steel bridge was built in 1904, and it is supported on the original masonry bases to the abutments on either side of the river. This stonework is all that remains of Brunel's ill-fated brick-built bridge.

Below: The iron girder bridge at the site of Brunel's Somerset Bridge. The stonework base of the abutments on either side is all that remains of the original.

Other Brick and Masonry Bridges
In addition to the many high-profile brick and masonry bridges covered, Brunel also built innumerable smaller bridges for his railways, mostly to cross minor obstacles or crossing points. All are workmanlike, as you would expect, but there are too many to list individually here. Instead a selection is presented overleaf. Note in particular the 'flying bridge' at Liskeard station in Cornwall. This masonry bridge springs from the sides of the cuttings. There is a similar example of this design with the 150-foot-wide Deals Bridge over the Bristol–Exeter line, a little to the south of Weston-super-Mare.

Above: The 'flying' arch at Liskeard station, in Cornwall, springs directly from the sides of the cutting to carry road traffic over the line. *Below:* A bridge over the cutting at Sydney Gardens, Bath. This uncluttered view is threatened by new overhead gantries when the line is electrified.

Three more representative examples of Brunel's many minor bridges on the railways.

Top right: The imposing entrance into the town through the substantial viaduct at Chippenham. The other side is of brick.

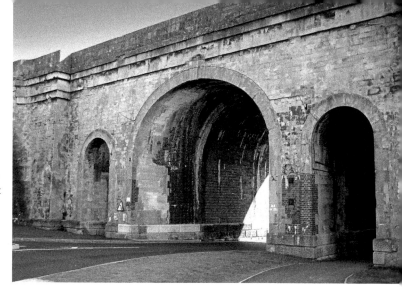

Middle: The road bridge immediately to the north of the station at Moreton-in-Marsh carries the A429 Fosse Way over the line.

Bottom: Built on a skew, a typical masonry bridge, this time over the B4060 road a little to the north of Wickwar station on the former Bristol & Gloucester Railway. This type of simple construction with dressed local stone is referred to as Ashlar.

Above: The main line crossed the Uxbridge Road at an awkward intersection. Brunel's first design was unsuccessful as it incorporated cast-iron girders supported by abutments and columns. *Below:* The bowstring bridge at Windsor used the far less brittle wrought iron. *(Laurie Lopes)* Originally the bridge was approached by low timber viaducts, since replaced.

Iron Works

Ever since Abraham Darby completed his famous cast-iron bridge at Ironbridge, a number of engineers have used this material in the construction of bridges. Cast iron is comparatively light and immensely strong in compression. Brunel designed three large cast-iron bridges for the GWR main line, the most ambitious being over the Uxbridge Road in Hanwell. This was at an awkward site where the railway crossing coincided with the junction of two turnpike roads which met at a very sharp angle. It was too sharp for a brick bridge and instead Brunel's design featured cast-iron beams supported on masonry abutments on either side, plus sixteen stout columns underneath. During construction of the bridge in 1837, one of the girders broke, and then another only a year after the bridge had opened. (Brunel also built a cast-iron road bridge to cross the canal at Paddington, covered at the end of this section.)

In 1847 the wooden decking on the Uxbridge railway bridge caught fire and, consequently, the ironwork was so badly damaged that the entire bridge had to be rebuilt. That same year, on 27 May, Robert Stephenson's cast-iron railway bridge over the River Dee, near Chester, collapsed as a train passed over it, resulting in the loss of five lives. This confirmed what the early Victorian engineers already suspected. Cast iron, so strong in compression, is brittle under tension. Wrought iron, on the other hand, is an iron alloy with a very low carbon content and has fibrous inclusions making it both strong and able to withstand bending forces. During the eighteenth century a number of processes for producing this malleable metal in large quantities had been devised, the most successful being the 'puddling' technique in which the molten iron was stirred within an enclosed puddling furnace without the use of charcoal. With further improvements in the manufacturing process wrought iron quickly became the preferred material for Brunel and his contemporaries. An additional advantage was that girders of considerable length could be formed by riveting wrought-iron sections together, a particularly useful attribute in building bridges.

Brunel gained considerable experience in working with wrought iron on the *Great Britain* steamship, and conducted a series of experiments into the strength of wrought-iron girders by testing different designs to destruction. From the results of this research he went on to create a number of significant bridges using the material, starting conventionally enough with the chains on the Hungerford suspension bridge over the Thames in London. More innovative designs for railway bridges soon followed. While the suspension bridge principle was suited to nineteenth-century road or pedestrian traffic it would not be able to carry a much heavier railway train, possibly weighing 100 tons or more, as it would sag under the weight, thus creating a ripple effect as the train passed along its length. As an alternative Brunel experimented with two 'bowstring' arch bridges, at Usk and on the line from Slough to Windsor.

All wrapped up, the iron bridge at Windsor undergoing renovation in the spring of 2014.

Bowstring Bridges at Usk and Windsor

The first, with a span of 100 feet, replaced a wooden span over the River Usk at Newport, and a bigger 203-foot-span bridge crossed the Thames at Windsor. This was on the Slough–Windsor branch line which was completed in 1849. This bridge featured three parallel bowstring girders – one for either side of the double track plus a central one – from which hung the vertical girders to support the roadway girders. Each semicircular rib of wrought iron has a maximum height of 23 feet and the bridge was positioned with a slight skew of 20 degrees. The ribs have a triangular cellular cross-section at the top which increases their strength, and the central rib is made from thicker plates. Wrought-iron girders contain the outward thrust or spreading of the three ribs and cross girders, and vertical hanger girders, plus cross-trusses linking the ribs, complete the structure.

Originally the Windsor bridge stood on cast-iron cylinders – much like those found at Chepstow – and it was approached via a long viaduct, originally of timber but later replaced by a brick structure, which crossed the flat meadows leading to the river. The bridge at Usk has since been replaced but Windsor bridge remains as the oldest surviving example of Brunel's iron bridges. It is readily accessible via the car park at the nearby sports centre, although in 2014 it was hidden by scaffolding and screens during a major renovation.

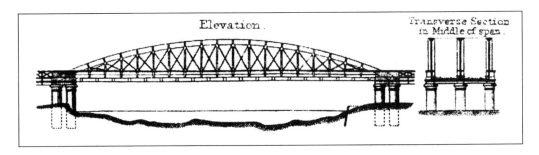

Bristol's Other Bridge

Brunel's next experiment with wrought-iron construction for bridges stemmed from his efforts to make improvements to the docks at Bristol. In the early eighteenth century Bristol had been second only to London as a port and the city's merchants had grown wealthy on the trade in tobacco, timber, sugar and cotton, but also through the appalling 'Africa trade' in slaves. However, by the nineteenth century the port was in decline, due largely to the tidal waters of the River Avon which restricted access to vessels through the winding course of the river, and twice daily left them high and dry. Accordingly a vessel had to be 'ship-shape and Bristol fashion', in other words it had to be well built if it was not to have its back broken on Bristol's treacherous mudbanks. In 1804 work commenced on William Jessop's scheme to create a Floating Harbour by enclosing the river in a section between the Neetham Dam, at Temple Meads, and the Rownham Dam on the western side of the city. The docks were kept permanently full of water, and vessels could enter or leave on the high tide via the locks at the Cumberland Basin. In addition the new Cut was created and this skirted around the south side of the docks to cope with the coming and going of the tidal waters. The Floating Harbour gave Bristol a much needed lease of life, but Jessop's scheme also brought its own problems. The River Frome, which fed fresh water into the docks, was insufficient to shift the mud, which began to clog up the harbour, and in 1832 Brunel was consulted on the matter. His solution was to improve the flow of water by installing gates to create an

An early postcard view of the entrance locks into the Cumberland Basin, showing the original and duplicate tubular bridges. Brunel's lock is on the right-hand side. The indentation between the two locks is for a gridiron.

Built by Brunel to cross his South Entrance Lock at Bristol, this little swing bridge is of particular interest as it utilises the tubular wrought-iron girder construction that will lead directly to his larger-scale railway tubular bridges – over the Wye at Chepstow and the River Tamar at Saltash. This historic bridge now stands beside the North Entrance Lock into the Cumberland Basin.

'underfall' which acted like the plughole in a bath. He also devised a scraper boat which was winched backwards and forwards to move the mud towards the entrance of the Cumberland Basin by means of a submerged blade.

His next measure was to make improvements to the South Entrance Lock between the river and the Cumberland Basin. Progress was slow and his new lock was eventually completed in 1849. It measured 262 feet by 54 feet and featured a semi-elliptical bottom for maximum clearance for laden vessels, and a new system of iron 'caisson' gates with internal chambers which contained a pocket of air to create partial buoyancy in high water. To allow road traffic to pass over the lock he devised a swing bridge with innovative tubular girders of wrought iron. It was to be the direct predecessor to the full-scale tubular bridges at Chepstow (1849) and the Royal Albert Bridge at Saltash (1859).

Unfortunately Brunel's South Entrance Lock did not remain in operation for very long. In 1873, the docks engineer Thomas Howard constructed the larger New Entrance Lock on the northern side of the harbour entrance. To save money Howard pinched Brunel's swing bridge for his lock as he considered the old lock to be redundant. When the ship operators complained, the city corporation agreed to keep the South Lock operational and they dusted off Brunel's drawings to create a replica bridge. This is the bridge that is fixed in position on Brunel's South Entrance Lock, while the original Brunellian wrought-iron bridge is now beside the north lock in the shadow of the modern A3029 Plimsoll swing bridge. Brunel's wrought-iron bridge represents an important stage in the evolution of his tubular bridge designs. Its survival over the years is, I suspect, more through the good fortune of neglect than any deliberate policy. The bridge is easily accessed on foot, although at the time of publishing (spring 2014) it was in need of restoration.

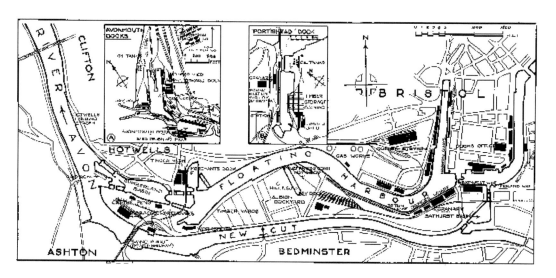

Lloyd's of London plan of the Bristol Docks from the 1930s. Brunel's South Entrance Lock is shown to the left of the Cumberland Basin. *(CMcC)*

65

Above: Brunel's South Entrance Lock into the Cumberland Basin at Bristol. This had a semi-elliptical base to allow better clearance for laden vessels, caisson-style gates that could be floated out of position, plus a swingbridge with innovative wrought-iron tubular girders. But that's not it in position on the lock, instead it is a later replacement and the original Brunel bridge, shown below in close-up, can be seen in the distance under the Plimsoll road bridge.

The Chepstow Railway Bridge

Brunel would take the principle of the tubular girder bridge and supersize it in order to carry the South Wales Railway over the River Wye at Chepstow. It had actually been his close friend, Robert Stephenson, who first applied the box girder principle on a large scale with his bridges for the Chester & Holyhead Railway at Conway, completed in 1848, and the Britannia Bridge over the Menai Straits in 1850. The Britannia Bridge, shown below, consisted of four spans of 460 feet, each in the form of huge box girders of wrought iron that were large enough to actually enclose the railway tracks within them.

The particular challenge Brunel faced at Chepstow lay in the asymmetrical nature of the site. There are 120-foot-high limestone cliffs on the eastern, Gloucestershire, side of the river and on the opposite Monmouthshire side there is an area of low ground comprising mostly shingle and clay which rises only slightly above the high water level. The Wye is about 600 feet across at this point. It also has a wide tidal range and the Admiralty was insisting that Brunel provided a clear span of 300 feet with a vertical clearance for its high-masted ships of at least 50 feet at high tide. This requirement ruled out the possibility of a supporting arch of any sort. Brunel's design for the Chepstow Bridge was unusual in that it combined the tubular girder principle with some aspects of a suspension bridge. On the Gloucestershire side of the bridge a 20-foot cutting led to a pair of towers pierced by arches rising about 50 feet above the rail decking. At the other end of the main span a similar pair of

Completed in 1850, Robert Stephenson's Britannia Railway Bridge across the Menai Straits was one of the first uses of tubular girders on any scale, in this case with the trains travelling inside the box girders. Note the openings at the top of the tower indicating that he had intended to incorporate elements of a suspension bridge in the design. *(CMcC)*

Opposite: A section of wrought-iron girder from the decking of the Chepstow bridge, revealing the internal triangular arrangement of the top flange. It is displayed at Brunel University in Uxbridge. *Above:* The lopsided nature of the site at Chepstow.

arches were supported on cylindrical iron columns, and beyond them were a further three 100-foot straight girder spans which met a raised abutment leading to an embankment. Between the pairs of towers on either end of the main span there were in effect two independent bridges side by side, each carrying a broad gauge track on a deck of plate girders. Suspension chains – adapted from the unused ones intended for the uncompleted Clifton bridge – curved downwards from the towers, while overhead two wrought-iron tubes 9 feet in diameter, reinforced with internal diaphragms at intervals, acted as box girders counteracting the compression forces. Vertical trusses or struts and diagonal ties linked the deck to the tubular girders to ensure rigidity, although the deck girders were not fixed to the chains, but sat on rollers and saddles. In engineering terms this type of bridge is referred to as a closed system as all of the compression and tension is contained within the structure, as opposed to the open system on a conventional suspension bridge where the loads are transferred by the chains to anchorage points on either side.

Chepstow tubular bridge is often cited as a dress rehearsal for the later and bigger Royal Albert Bridge at Saltash. This might be a fair observation in the broader picture of Brunel's career, but we should not underestimate the significance of the Chepstow bridge simply because it is no longer there. It was at Chepstow that Brunel honed the techniques that would serve him so well at Saltash. These included the use of caissons, air-filled iron chambers

Above: Brunel's bridge over the Wye at Chepstow. This is a 'closed' system with the forces of compression and tension contained within the structure, a design necessitated by the site, with high cliffs on one side and an area of open ground on the other. In effect it was constructed as two bridges side by side. Note that the tubular girders are arched very slightly, a fact missed in many contemporary illustrations. *Below:* An interesting postcard of the *Westward Ho* paddle steamer which includes the bridge almost incidentally. *(CMcC)*

The Chepstow site as it is now. By the 1950s the old bridge had become weakened and in 1962 it was replaced by this underslung girder truss. The A48 road bridge, nearer to the camera, obstructs the view of the rail bridge. *Below:* The original support columns still remain.

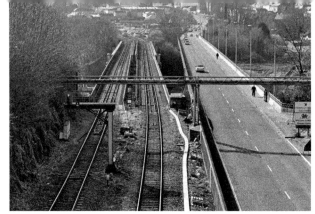

Left: Looking down on the Chepstow crossings with the rail bridge on the left and the A48 road bridge on the right. You can walk across the road bridge and it does provide a great vantage point from which to view the rail bridge, as shown below, but do note that there is no footpath on that side of this dangerous road.

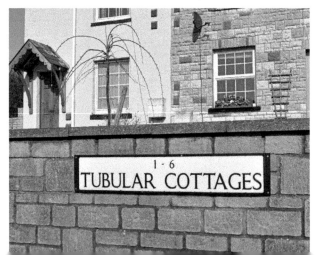

Left: An unexpected reminder of the Chepstow Railway Bridge, these appropriately named cottages on the edge of Chepstow are on a quiet suburban road which passes over the approach on the Gloucestershire side. Such road names can often provide pointers to an area's history.

to protect the workers building the iron support piers, and also the method of constructing the tubular girders on the shore and then floating them into position. On 8 April 1852 the first of Chepstow's 460-ton tubular girders was placed onto a pontoon consisting of six wrought-iron barges. Taking advantage of the high spring tide it was guided into position on the river, then lifted up to the level of the railway, and afterwards to its place between the towers. The first line of track was operational by July 1852 and the other was completed shortly afterwards the following year. *The Illustrated London News* observed:

> The peculiarity of the site did not permit any display of Art – that is, of architectural embellishment; indeed, a pure taste rejects any attempt to decorate a large mechanical work with sham columns, pilasters, and small ornaments.

It appears that the Victorians weren't quite ready for the aesthetics of pure functionalism, although in truth the subtle shaping of the towers has something of the Egyptian flavour already displayed by Brunel on the Clifton Suspension Bridge and on the piers of the Wharncliffe Viaduct across the Brent Valley at Hanwell. This understated Egyptian influence was also carried over to the towers on the Royal Albert Bridge at Saltash, which is rightly regarded as Brunel's final masterpiece. By the 1950s the Chepstow bridge was becoming too weak for the increasingly heavy rail traffic and a speed limit of 15 mph was introduced. In 1962 Brunel's original tubular structure was dismantled and an ugly underslung truss now supports the tracks.

Below: Dismantling Brunel's tubular bridge at Chepstow in 1962. The underslung girder truss is already in position. Note the suspension chains on the left.

The Royal Albert Bridge, Saltash

This bridge has been described as a 'masterpiece of complexity' and it is easy to see why. As with Chepstow's bridge, it is a closed structure incorporating three engineering forms: the compression arch, the tension chains of a suspension bridge, and a beam deck. In the photographs it is easy to make out the suspension chains which mirror the twin humps of the tubular girders. Look again at the towers, and you will see how closely they resemble those of the Clifton Suspension Bridge, and the connection goes further than that as some of the chains came from Bristol when that bridge was left unfinished. However, the big difference is that at Bristol the chains pass over the towers and are anchored into the rock on either side creating an 'open' structure where the forces are equalised. With the double span at Saltash there is nowhere for the chains to go and so their tension forces, in effect pulling the towers towards each other, are countered by the tubular arches at the top. It is a bowstring design similar in some respects to the bridge at Windsor, but with an extra twist in the design in the form of the chains. Furthermore, the arched girder is much more substantial, and to further stiffen the structure Brunel incorporated vertical hangers or struts running from the deck to the arches, with additional diagonals for longitudinal stability.

Brunel's involvement with the railway into Cornwall had begun in February 1845 when he appeared before a parliamentary committee examining a Bill submitted by the Cornwall Railway. He was already engineer to the South Devon Railway, and the plan was to extend the existing coastal line from Devon with the backing of the so-called Associated Companies – mainly the GWR

Opposite: The main pier on the Saltash side of the bridge. *Below:* An engraving of the Royal Albert Bridge long before the arrival of the later road bridge alongside.

Above: The second truss in position on the Devon side. It will be raised up, little by little, as the masonry pier is completed. Note that the tubular girders are oval, not circular, in cross section.

and other broad gauge companies. In isolation a Cornish railway made little financial sense, but the existing companies would reap the benefits of carrying additional passengers for Cornwall over their lines. The only problem was getting across the River Tamar.

One plan was to load the trains onto special ferries, which would have proved a very awkward procedure as the fast-flowing waters of this part of the Tamar have a rise and fall of 18 feet. Brunel even gave his tacit approval of the scheme, as he said to the parliamentary committee, 'I am prepared to say that I consider there is no difficulty in doing it.' As even Brunel must have anticipated the ferry plan fell through and the directors of the railway

went away to reconsider their options. Spurred on by a rival plan to take the line further inland via Oakhampton, they appointed Brunel as engineer in August 1845. He immediately started a new survey of the route, including a bridge at Saltash, and by the time he reappeared before a parliamentary committee the rival inland scheme had fizzled out. With Brunel on board the Cornwall Railway Act received royal assent in August 1846. However, the exact form of the bridge over the 1,000-foot-wide stretch of the Tamar had yet to be determined. Brunel had considered a massive single span, possibly of timber, because as with Saltash the Admiralty was making stipulations regarding clearance for its tall-masted ships. But within a year the first flush of railway growth, 'railway mania' they called it, had run its course and there was a slump in the value of railway company shares. This allowed Brunel to pursue his many other projects, and also gave him time to work on his ideas for the Tamar. Crucially, it was time well spent in building the tubular bridge at Chepstow.

When Brunel came back to the Saltash project, he settled upon a design featuring two identical spans of 465 feet, although this was reduced to 455 feet later on, with a single mid-river pier. On dry land on either side would be a main pier linked by a series of girder spans carried on several smaller piers. The biggest challenge was the construction of that central pier and in

A view of the truss taken from the road bridge. The horizontal bracing is a later addition.

order to investigate the nature of the riverbed, in 1848 he had a wrought-iron tube constructed, 85 feet long and 6 feet in diameter. Acting like a cofferdam, this tube was lowered with one end resting on the riverbed and pumped out to enable the drilling of a number of trial borings. Fortunately this revealed a bed of hard rock beneath a thick layer of mud and slime. Armed with this information Brunel was able to further his designs, but once again a lack of funding meant that matters were put on hold for three years. In 1851 he attempted to resuscitate the Cornwall Railway project by proposing that savings could be made by reducing the line from Plymouth to a single track, including that on the Saltash bridge. The latter was approved in 1852, by which time Brunel's designs were finalised in the form we see today.

The first task in building the bridge was to erect the central pier. The main part, visible above the waterline, consists of four octagonal iron columns 96 feet high, but beneath them is a masonry column descending a further 96 feet into the water. To construct this, Brunel built another cylinder, a much bigger one at 35 feet wide. In effect this functioned like a diving bell in which the stonemasons could work. To keep the river water from seeping in, the air within the cylinder was kept under pressure by steam-driven pumps, and this made it possible for the men to work 70 feet beneath the water, although they frequently suffered the severe headaches and muscular pains we now associate

Opposite: The central pier consists of four hexagonal columns in wrought iron. A masonry column of similar height descends to the riverbed. *Below:* An image of the bridge, *c.* 1890s.

Seen from Saltash station:

Left: The opening of the bridge in May 1859. By this time Brunel had become too ill to attend the ceremony. *Below:* Photographed in 2005. Note how the double tracks converge into one.

Left: In 1892 the last of Brunel's broad gauge track was ripped up and replaced with the so-called standard or 'northern' gauge which the Stephensons had adopted on their early railways. The nation needed a standard and their gauge won by sheer weight of miles covered.

with the 'bends'. The central pier was completed by 1856 and the great cylinder was unbolted, split into two and removed. In comparison the construction of the two main side piers, the portal piers, and the seventeen narrower piers on the approach spans – eight for Devon and nine on the Cornwall side – was a far simpler matter. The portal towers have a masonry and brick lining within an iron shell. The equivalent portal on the central pier is hollow and made of wrought iron.

While work progressed on the various piers, the first of the two trusses was taking shape on the Devon shore. The curving upper girders were oval in cross section being 12 feet 3 inches high and 16 feet 9 inches across. They are hollow and it is possible for a workman to pass through their entire length. The oval shape served the dual purpose of reducing sideways wind resistance and also permitting the deck struts to hang vertically, whereas those at Chepstow were inclined. Each truss weighed 1,000 tons and was to be positioned at the bottom of its piers by floating it into position on hugh pontoons which could be raised or lowered in the water by pumping water in or out of them. The day for moving the first truss to the Cornwall side of the bridge was set for

Below: The artist Terence Cuneo was famed for his railway posters. This is his evocative cover artwork for the 1963 Triang model railway catalogue.

A fascinating set of images from the GWR's 1935 publication *Track Topics*, subtitled as 'A Book of Railway Engineering for Boys of All Ages'. It shows the replacement of the approach decking and also the inside of a tube. Note the chap walking across the top with nothing more than a low rail to hook a foot under.

1 September 1857. Brunel made meticulous plans for the operation and on the day he took command from a platform mounted high up on the truss, accompanied by a team of flag-waving signallers to convey his instructions to the various teams of workmen. Thousands of onlookers, some estimates suggest 300,000, crammed the shoreline and hills to witness this wonder, but they had been warned that absolute silence was demanded by the great engineer. As the signal flags fluttered out their commands, one eye-witness recorded the event:

> Not a voice was heard, not a direction spoken; a few flags waved, a few boards with numbers on them were exhibited, and, by some mysterious agency, the tube and rail borne on pontoons travelled to their resting-place, and with such quietude as marked the building of Solomon's temple. With the impressive silence which is the highest evidence of power, it slid, as it were, into position without an incident, without any extraordinary effort, without a 'misfit' to the extent of the eighth of an inch. Equally without haste and without delay, just as the tide reached its limit, at 3 o'clock, the tube was fixed on the piers 30 feet above high water.

This was Brunel's moment. He was the master magician, in complete charge as he reshaped the landscape to his bidding. When the band struck up 'See the Conquering Hero Comes', the spell of silence was broken and the vast crowd erupted into a cacophony of cheering.

The process of raising the truss to its final position was a far more protracted matter, with huge hydraulic jacks lifting it 3 feet at a time as the iron sections were added to the central pier and the masonry was put in place on the land pier. The mortar had to be given time to fully set between each lift and it wasn't until May the following year, 1858, that it was finally in position. By the time that the second span, the one on the Devon side, was ready to be moved Brunel was in London struggling with the *Great Eastern* steamship. Accordingly his most trusted assistant, Robert Brereton, supervised the floating of the span and its positioning in July 1858, and it was under his guidance that the bridge was completed in early 1859. By this time Brunel's health was in decline as he was suffering from a kidney condition known at the time as Bright's disease. When Prince Albert travelled by GWR train down from Paddington to open the Royal Albert Bridge, as it was to be known, in May of 1859, Brunel was too ill to attend the ceremony. Shortly afterwards he saw the completed bridge for the first and last time. Reclining on a couch placed on an open trolley, he was drawn slowly across the bridge by one of Daniel Gooch's steam locomotives. It was a poignant moment. 'No flags flew, no bands played, no crowds cheered ...'

He died in London just a few months later on 15 September 1859. As a tribute the directors of the Cornwall Railway had the following epitaph added to the portals at each end of the bridge: 'I. K. BRUNEL ENGINEER 1859.'

Opposite: The 'flying banana' assesses the condition of the track. *(Network Rail)*

Above: The Royal Albert Bridge undergoing a major renovation, photographed in 2013 looking towards Saltash on the Cornish side. The huge tent-like structures enabled the work to carry on continuously while preventing debris falling into the river.
Left: Just as with the Clifton Bridge, the RAB has become an iconic symbol. This sign is on the Saltash station platform.
Below: The Crathie or Balmoral Bridge, an iron bridge built in 1856 for the royal estate in Scotland. *(Andy Kelly)*

Today the Royal Albert Bridge at Saltash is little changed and remains in constant use, although even the sleek HST 125s expresses are humbled by a 15 mph speed limit while crossing. The broad gauge track was removed in 1892 and there have been a few minor modifications including the addition of horizontal girders to strengthen the main spans, and the iron approaches have been replaced by steel. More recently a £10 million refurbishment of the bridge has been completed. Work had started in 2010 and the engineers would spend nearly two million hours of work strengthening and repainting the bridge. The latter process has thrown up an interesting discovery. By stripping away countless layers of paint the engineers have revealed that when completed in 1859 the spans had been finished with a pale stone or off-white colour. That didn't last long though, and red-brown anti-rust paint was slapped on within a decade, along with the 'goose grey' which we are more familiar with today. The refurbishment has been carried out concurrently from each end of the bridge, and to provide a safe working environment, as well as to contain any dust and debris from falling into the river, the sections being worked on have been encapsulated in tent-like structures.

The most conspicuous addition since Brunel's day is the A38 Tamar Road Bridge, a suspension bridge as it happens, which opened in 1961. Although it clutters the view from some angles, it does provide a superb viewing platform from which to see the old bridge. There is a pathway for pedestrians and cyclists, but note that cars going into Devon pay a toll.

The Hidden Bridge and a Royal Connection

There are a couple of interesting footnotes to Brunel's iron bridges. In 2004, Steven Brindle of English Heritage was examining Brunel's notebooks as part of his research on Paddington station when he came across records for load-testing the iron beams of a canal bridge dating from 1838. Correspondence was also found from Brunel to the Grand Junction Canal Company. Nobody knew if the bridge still existed beneath the later additional structures. Then by good luck it was discovered beneath the brickwork of the Bishop's Road Bridge, which was in the process of being replaced. The railings and upper structure had gone, but incredibly the iron arches, which were infilled with soffit plates, were still intact. They were removed for safe-keeping at an English heritage storage facility, and it is hoped that a new site can be found for what is the oldest surviving example of Brunellian cast-iron bridge.

The royal connection comes courtesy of Prince Albert who commissioned Brunel to design a bridge over the River Dee at Crathie on Queen Victoria's Balmoral Estate in Scotland. Built in 1856, and sometimes referred to as the Balmoral Bridge, it is a girder bridge of simple, almost austere, design. Two 130 -foot wrought-iron girders are supported on 18-foot-high masonry abutments. The parapets are decorated by a pattern of diagonal webplates. The ironwork for this minor bridge was made by R. Brotherhood of Chippenham, Wilts.

Timber

Although revered as a man of iron, Brunel was not afraid to use other materials. This included timber when circumstances dictated, which meant, more often than not, when the budget was tight, and not just for his railway buildings but also for more substantial structures such as bridges and viaducts.

In 1841, Brunel constructed a wooden bridge to carry a public road over the railway cutting at Sonning on the eastern side of Reading. This 2-mile swathe cut through the landscape at depths varying from 20 feet to nearly 60 feet and at the level of the road crossing it was 240 feet wide. Once again J. C. Bourne provides us with an excellent depiction of the bridge, which stood on four timber piers or trestles, two on either side of the track and another shorter pair halfway up each slope. The roadway consisted of a timber platform carried on three longitudinal beams which were supported by timber struts radiating like a fan from the top of the piers at about 12 feet below the road. There was a twofold advantage to this method of construction using timber – it was relatively cheap as well as being quick to build – and Brunel would revisit this configuration for other timber viaducts, most notably in Devon and Cornwall.

Further down the line Brunel devised his first timber bridge to carry the railway itself, this time at an awkward skew angle over the River Avon immediately to the west of Bath station. As we have seen, Brunel mistrusted cast iron as a material for bridges, and under pressure from the Bristol Directors of the GWR to curb his overspending he came up with a unique

Opposite: Detail of the wooden structure on the Carvedras Viaduct in Truro, Cornwall. Note how the timbers fan out from the top of the masonry pier. *Below:* This road crossing over the Sonning Cutting is one of the earliest examples of Brunel's use of timber for a bridge.

The river crossing immediately to the west of Bath station was at too sharp an angle for a brick bridge and Brunel devised this timber structure consisting of six timber ribs. It has masonry abutments and central pier, and the decorative infill is of iron. The abutments and pier remain but the wooden bridge has been replaced by a conventional lattice truss girder.

design for a timber bridge. This had two spans, each consisting of six laminated ribs, built up of layers of wood bolted together, resting on masonry abutments and a single central pier. Iron ties connected the ends of the wooden ribs and the inner spandrels, the spaces between the arches and the platform, were criss-crossed by iron ties and braces. The open spaces on the outer ribs were filled in by ornamental cast ironwork. Bath's distinctive skew bridge was replaced by an ugly steel girder bridge in 1878, although the original masonry components still remain.

As the branch lines began to proliferate during the 1840s, Brunel emulated his design for the Sonning road bridge on the construction of a number of wooden viaducts on the Swindon–Gloucester line, including nine in the 7-mile stretch between Frampton and Stroud alone. For the Bourne viaduct, built across the Stroudwater Canal in 1842, the main span of 66 feet was supported by triangular trusses which rested in iron shoes on the top of masonry piers. On the St Mary's Viaduct the main span was wider still at 74 feet. That same year a timber viaduct was completed on the Bristol & Gloucester Railway at Stonehouse and this consisted of five timber trusses, 50 feet wide, standing on timber trestles. He also designed timber bridges for other railways including the Oxford, Worcester & Wolverhampton Railway, and the Birmingham and Oxford Junction.

Brunel built upon his experience by conducting a series of experiments to ascertain the strength of larger timbers and by exploring techniques for preserving the wood, in particular the Kyanising method which involved soaking the pre-cut timbers with perchloride of Mercury. This led to a far greater use of the material as the railway network spread across southern Wales and, later on, south-westwards into Devon and then Cornwall. The

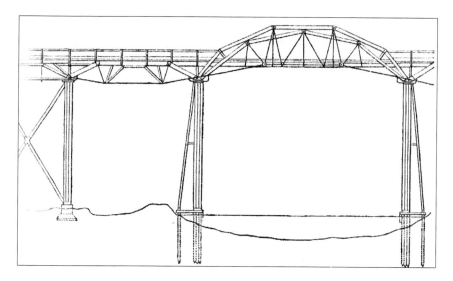

The longest of Brunel's timber structures was the viaduct at Landore, crossing the Swansea Valley in South Wales. 1,760 feet long it had a central span of 100 feet.

This engraving was published in *The Engineer* in June 1892, just a month after the demise of the broad gauge, in an article about the replacement of the timber viaducts within Devon. It also serves to illustrate the different types of viaducts used in the county, in particular the way the timber supports fan out from ledges near the top of the piers and are integrated into the balustrade. Note in particular the image of Monksmoor Bridge, top right, which shows a 'King' truss as commonly used on smaller bridges. The image top left shows the construction of the new Ivybridge Viaduct beside the old one. *Below:* The curving sweep of the Moorswater Viaduct in Cornwall. One of the most spectacular of Brunel's timber viaducts, it was 145 feet high and 954 feet long with fourteen pierced and buttressed piers.

Above: A dramatic setting for the viaduct at St Austell in Cornwall, with the original piers standing tall beside the masonry viaduct which replaced the timber one in 1898.

longest of Brunel's timber viaducts was at Landore, near Swansea. 1,760 feet long it had thirty-seven openings with a variety of spans ranging from the longest at 100 feet and down in increments to the smallest of around 40 feet. The piers were of different materials, either masonry, timber or a combination of both, depending on the nature of the foundations. The main central truss consisted of a double polygonal frame connected by bolts and struts, and further strengthened by wrought-iron tie-bars. A similar design was used at Newport, where there were only eleven spans, but this was destroyed by a fire before it had been finished and it was replaced by wrought-iron trusses for the main span.

Back in England, the South Devon Railway had to pass over four deep valleys between Totnes and Plympton on the outskirts of Dartmoor. Four timber viaducts were constructed along the same design, the largest being at Ivybridge. This was on a curve and had eleven spans of 61 feet, standing at its highest point at 104 feet above the valley floor. Designed to carry the atmospheric railway initially, the timber decking and framework were supported on pairs of slim masonry columns. When locomotives were introduced to this stretch of the line the viaduct was strengthened with an additional trussed parapet above the existing trusses.

Such was Brunel's confidence in timber construction that he even proposed a timber bridge for the difficult crossing over the Tamar at Saltash. If this wooden wonder had ever been built it would have been the greatest timber bridge the world had ever seen,

93

One of the last to go, the Ponsanooth Viaduct on the Falmouth branch line, with the replacement nearing completion in 1930. As in many cases this was built alongside the old viaduct.

with six spans of 100 feet and a central one of 250 feet. It is possible that IKB had taken his inspiration from his father, Marc, who forty years previously had proposed a 'great bridge' with an 800-foot laminated timber arch to cross the River Neva at St Petersburg. Fortunately for posterity's sake, if nothing else, IKB turned to wrought iron to create what became one of his finest works, the Royal Albert Bridge.

Work on the Cornwall Railway began in 1852, even though the Royal Albert Bridge wouldn't be completed for another seven years. It was particularly challenging countryside with a large number of deep valleys to be crossed requiring the construction of forty-two viaducts between Plymouth and Falmouth. For the most part these consist of masonry piers 60 feet apart, centre to centre, rising up to within 35 feet of the decking. The piers were capped with iron plates from which four sets of timber struts radiated like an upturned hand, with further cross-bracing to support longitudinal beams beneath the deck. The tallest of this type of viaduct stood 153 feet above the ground. On the West Cornwall Railway, where the height was not so great, the spans were at 50-foot intervals and sat on timber piers. There were other variations to the structures. For example, in a shallow valley – such as the one at Pendalake to the east of Bodmin Road station – the deck was supported on inclined timber legs sitting on small masonry piers. These piers were situated at 40-foot intervals and rose only slightly above groundlevel. On the St Germans Viaduct the span was also 40 feet with support from a tapered timber and wrought-iron truss.

On the whole, the design of the timber components was standardised, partly to keep costs down, partly to facilitate easy replacement. Inevitably the timber would not last forever and during the lifetime of the viaducts special maintenance teams became adept at replacing individual struts. Because the price of Baltic pine continued to rise steadily, by the time the main-line track was doubled in 1908 the viaducts had all been replaced (a few of the timber viaducts lasted on the branch lines until the 1930s). In some cases the timber components were removed and the masonry piers were extended up to the full height of the deck. There is a good example of one of these at Menheniot, just 8 miles into Cornwall along the A38 near Lower Clicker. Others did not fare so well and entire new viaducts were built alongside the old ones, leaving the unwanted piers standing like a row of tombstones.

Brunel has been described as 'the greatest timber engineer Britain has ever seen', but sadly not a single one of his timber bridges has survived.

St Germans Viaduct. 945 feet long, it was completely replaced by this masonry one in 1907.

Further reading:

Also by John Christopher:
Brunel in London (Amberley Publishing, 2013).
Brunel in Bristol (Amberley Publishing, 2013).
The Lost Works of Isambard Kingdom Brunel (Amberley Publishing, 2011).

Brunel – In Love With the Impossible, Edited by Andrew and Melanie Kelly (Brunel 200, 2006).
Brunel – The Man Who Built the World, Steven Brindle (Weidenfeld & Nicholson, 2005).
Isambard Kingdom Brunel, L. T. C. Rolt (Longmans Green, 1957).
Isambard Kingdom Brunel – Engineering Knight Errant, Adrian Vaughan (John Murray, 1991).
Isambard Kingdom Brunel – Recent Works, Edited by Eric Kentley, Angie Hudson and James Peto (Design Museum, 2000).
The Life of Isambard Kingdom Brunel – Civil Engineer, Isambard Brunel (Longmans Green, 1870).

Acknowledgements:

I would like to thank the following individuals and organisations for providing photographs and other images for this book: Campbell McCutcheon *(CMcC)*, the US Library of Congress *(LoC)*, Network Rail, Andy Kelly and Laurie Lopes. Unless otherwise indicated all new photography is by the author. JC